POSTAL
STATION
FROM

FROM
TO

POSTAL
STATION
TWO CENTS
FROM

FROM

POSTAL
STATION
FROM

POSTAL
STATION
FROM

FROM

FROM

FROM

POSTAL
STATION
FROM

POSTAL
STATION
FROM

SOUTH
POSTAL
STATION
FROM

FROM

SOUTH POSTAL STATION
FROM

POSTAL
FROM

FROM

FROM
TO

POSTAL
STATION
FROM

FROM
TO

SOUTH POSTAL STATION
FROM

FROM

FROM

SOUTH
POSTAL
STATION
FROM

FROM

POSTAL
STATION
FROM

SOUTH
POSTAL
STATION
FROM

FROM

FROM

FROM

SOUTH POSTAL STATION
FROM

FROM

POSTAL
STATION
FROM

FROM

POSTAL
STATION
FROM

FROM

FROM

FROM

POSTAL
STATION
FOREIGN TWO CENTS
FROM

FROM
TO

FROM

FROM

FROM

FROM

POSTAL
STATION
FROM

FROM

FROM

FROM

FROM

FROM

FROM

FROM

POSTAL
STATION
FROM

FROM

POSTAL
STATION
FROM

FROM

FROM

FROM

POSTAL
STATION
FROM

FROM

POSTAL
STATION
FROM
TO

POSTAL
STATION
FROM

FROM

POSTAL
STATION
TWO CENTS
FROM

FROM
TO

POSTAL STATION
FROM

FROM

POSTAL
STATION
FROM

POSTAL
STATION
FROM

POSTAL
STATION
FROM

POSTAL
STATION
FROM

www.ingramcontent.com/pod-product-compliance
Lightning Source LLC
LaVergne TN
LVHW080558200726
843510LV00004B/950
* 9 7 8 3 3 4 7 0 2 5 8 2 0 *